AF575499

Ocean Animals
Orcas
by Betsy Rathburn
BELLWETHER MEDIA
MINNEAPOLIS, MN
BLASTOFF!
Beginners

Blastoff! Beginners are developed by literacy experts and educators to meet the needs of early readers. These engaging informational texts support young children as they begin reading about their world. Through simple language and high frequency words paired with crisp, colorful photos, Blastoff! Beginners launch young readers into the universe of independent reading.

Sight Words in This Book

an	each	not	these	white
and	have	other	they	who
are	help	see	this	
big	in	the	time	
black	is	them	to	

This edition first published in 2022 by Bellwether Media, Inc.

Library of Congress Cataloging-in-Publication Data

Names: Rathburn, Betsy, author.
Title: Orcas / by Betsy Rathburn.
Description: Minneapolis, MN : Bellwether Media, Inc., 2022. | Series: Blastoff! beginners: ocean animals | Includes bibliographical references and index. | Audience: Ages 4-7 | Audience: Grades K-1
Identifiers: LCCN 2021001461 (print) | LCCN 2021001462 (ebook) | ISBN9781644874813 (library binding) | ISBN 9781648343896 (ebook)
Subjects: LCSH: Killer whale--Juvenile literature.
Classification: LCC QL737.C432 R38 2022 (print) | LCC QL737.C432 (ebook) | DDC 599.53/6--dc23
LC record available at https://lccn.loc.gov/2021001461
LC ebook record available at https://lccn.loc.gov/2021001462

Editor: Amy McDonald Designer: Laura Sowers

Printed in the United States of America, North Mankato, MN.

Table of Contents

Orcas!

Who is black
and white
and big?
An orca!

Orcas are not fish.
They are dolphins!

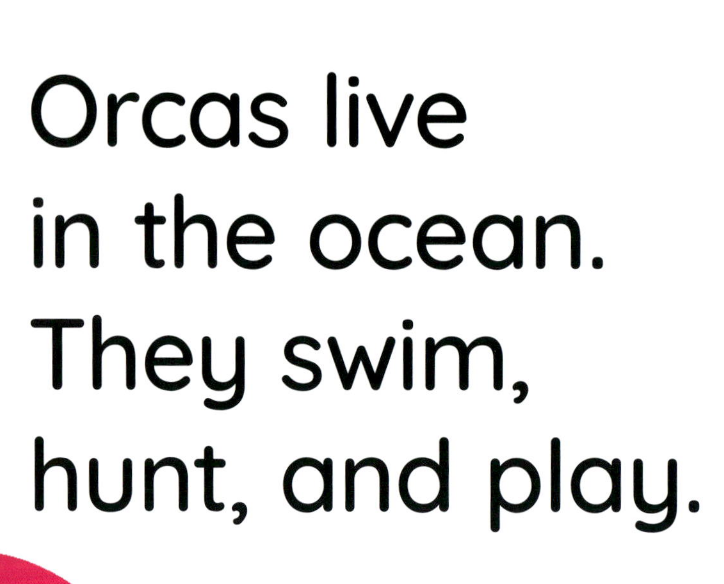

Orcas live
in the ocean.
They swim,
hunt, and play.

Body Parts

Orcas have **fins**. These help orcas swim fast!

fins

Orcas have
big teeth.
They bite food.

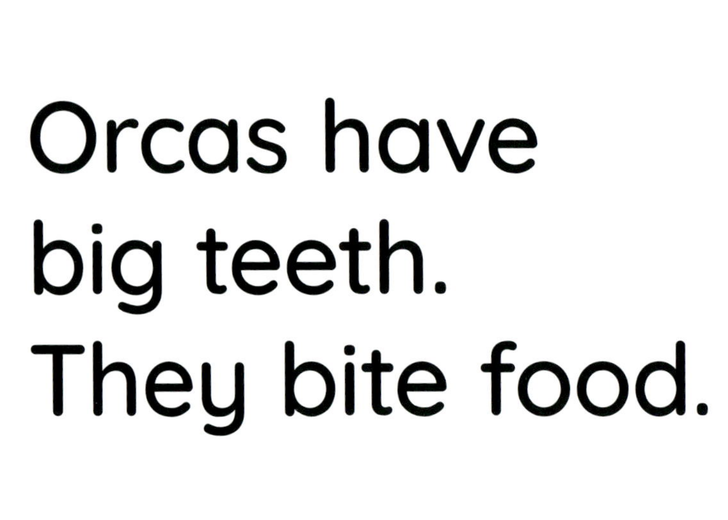

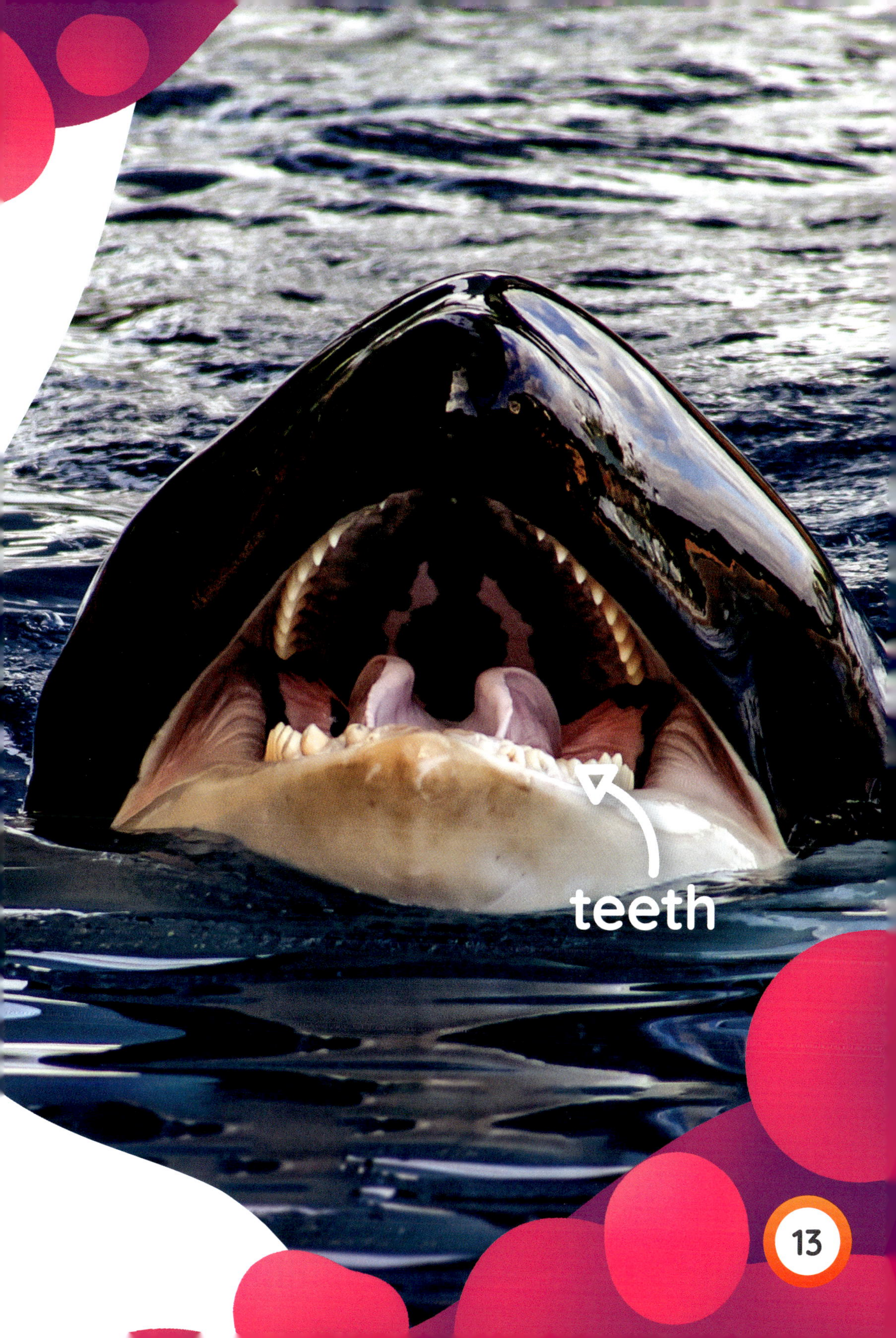
teeth

Orcas have **blowholes**. These let them breathe.

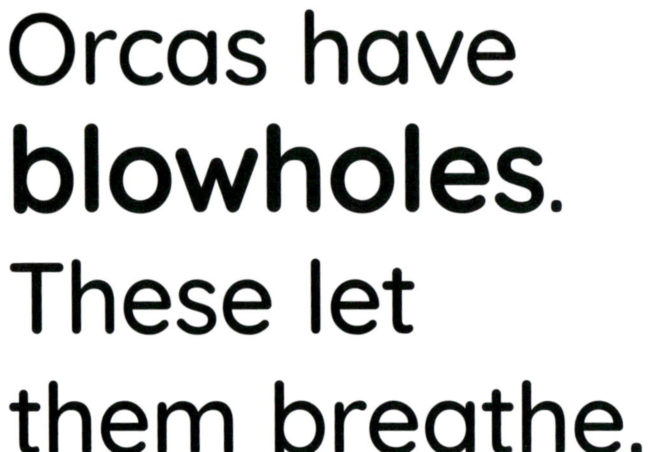

Pod Pals

Orcas swim
in **pods**.
They stay
together.

pod

Orcas talk
to each other.
They hunt
seals together.

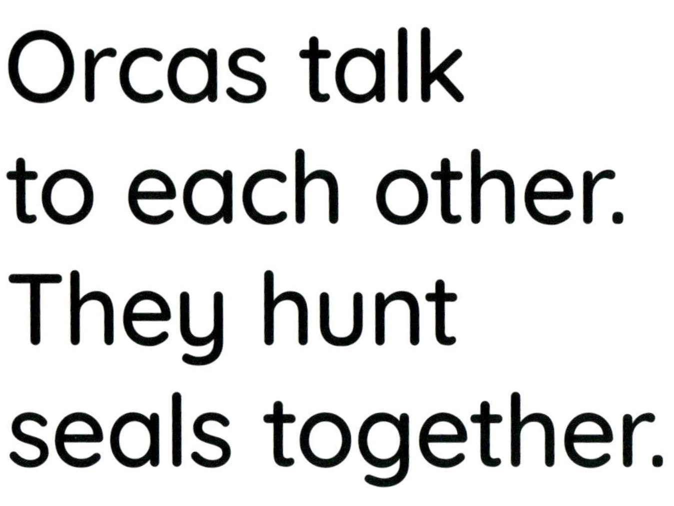

This orca sees food. Lunch time!

Orca Facts

Orca Parts

Orca Food

seals

fish

penguins

Glossary

blowholes

holes that orcas use to breathe

fins

flat body parts on an orca that help it swim

pods

groups of orcas

To Learn More

ON THE WEB

FACTSURFER

Factsurfer.com gives you a safe, fun way to find more information.

1. Go to www.factsurfer.com.
2. Enter "orcas" into the search box and click .
3. Select your book cover to see a list of related content.

Index

The images in this book are reproduced through the courtesy of: sakiflower1988, front cover; Wirestock Creators, p. 3; Love Lego, pp. 4-5; Nick Grobler, pp. 6-7; Doptis, pp. 8-9; Janik Rybicka, pp. 10, 23 (fins); Nature Picture Library/ Alamy, pp. 10-11; Guillermo El Oso, pp. 12-13; Christian Musat, pp. 14, 22; Design Pics Inc/ Alamy, pp. 14-15; Willyam Bradberry, pp. 16-17; Eric Isselee, p. 18; Sue Clark/ Alamy, pp. 18-19; Vladimir Seliverstov/ Alamy, pp. 20-21; Erwim Niemand, p. 22 (seals); DJ Mattaar, p. 22 (fish); fieldwork, p. 22 (penguins); Frederik Jegsen, p. 23 (blowholes); Tory Kallman, p. 23 (pods).